Easy-Step Books

Plumbing

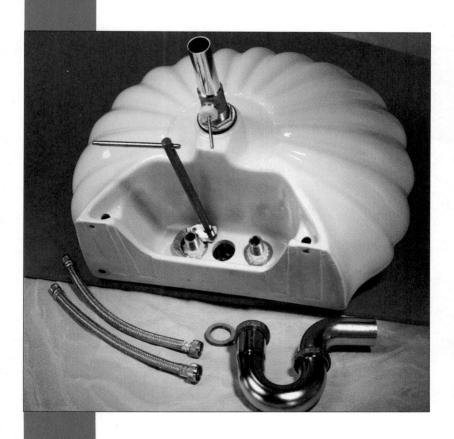

Contents

Plumbers are often called in to make simple repairs that homeowners could do themselves—and save a lot of money in the bargain. Why do people avoid plumbing chores?

One reason is that many find plumbing a bit mystifying. It needn't be. A peek through the walls of your home shows residential plumbing to be composed of two basic systems: the water supply system and the drain-waste-vent (DWV) system. The supply system carries hot and cold water to the faucets, fixtures, and water-using appliances of the home. The DWV system carries the waste from all the fixtures in the house to a municipal sewer line or septic tank. The vent pipes provide air to the DWV system, allowing waste to drain freely. They also provide a means of venting harmful sewer gas. Most plumbing problems have to do with a minor or major failure in one of these systems.

Plumbing is your one-stop guide to basic plumbing projects. You may be amazed at how quickly you'll gain the confidence to advance from fixing a leaking pipe to replacing an old sink. For best results, read through the entire book before you begin any work: You'll be less likely to make mistakes and should find your task easier.

Although this book presents a broad range of do-it-yourself repairs and improvements, it does not cover gas and fuel-oil piping. Both should be left to the experts. Plumbing and electricity meet at several places in the house (the dishwasher, garbage disposer, and washing machine, for example). Be sure to turn off the electric power before you begin plumbing work in these areas. You should be able to handle many plumbing repairs and improvements on such appliances, but leave the electrical work to a professional.

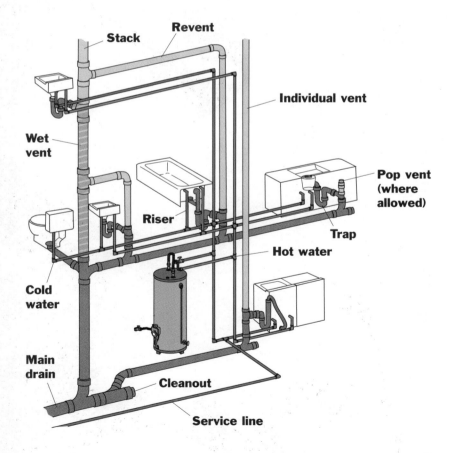

Faucets

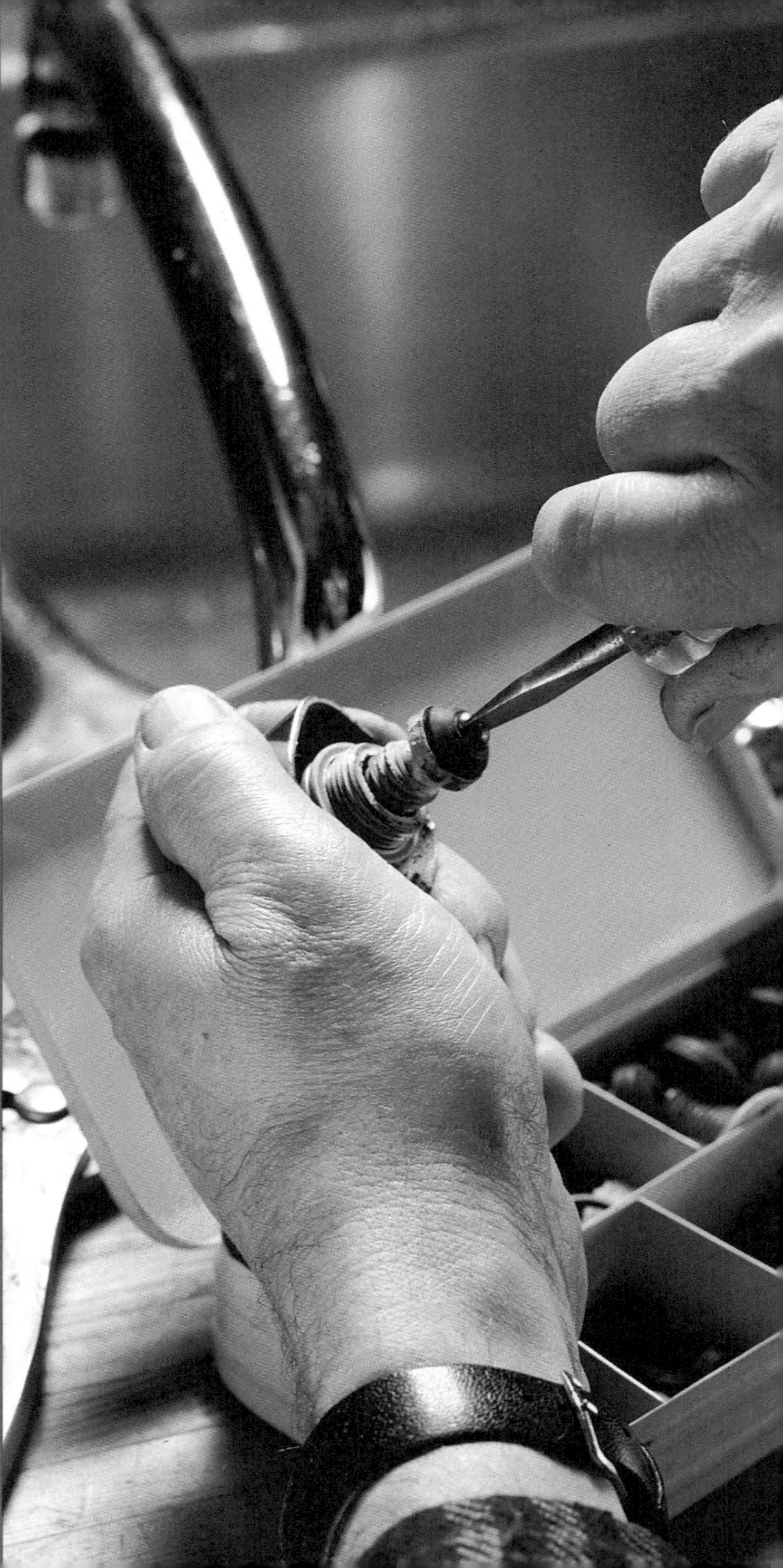

1 Improve flow

First make sure the shutoff valves are completely open. Next unscrew and disassemble the aerator, noting how the parts go together. Soak the parts overnight in a vinegar-and-water solution. Scrub with an old toothbrush and replace worn parts. Reassemble the aerator parts in proper order. If the problem persists, turn off the water supply and remove the supply risers. Check for clogs at all connections. Clean risers with wire or a long screwdriver. Continued flow restriction indicates clogged shutoff valves or pipes, or undersized pipes.

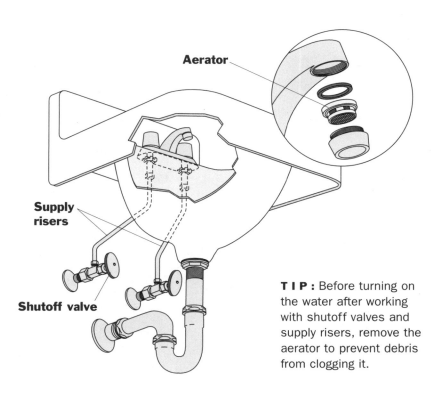

Aerator

Supply risers

Shutoff valve

TIP : Before turning on the water after working with shutoff valves and supply risers, remove the aerator to prevent debris from clogging it.

2

Fix compression faucet

Turn off the water supply to the faucet. Remove the handle screw, which may be under a decorative cap, then lift off the handle. If the handle is stuck, use a handle puller. Unscrew the retaining nut and lift out the stem. Remove the stem screw, washer, and O-ring and replace each part with an exact duplicate. Coat all of the parts with heat-proof grease, specially formulated for faucets. If the stem is badly damaged, replace it or install a new faucet.

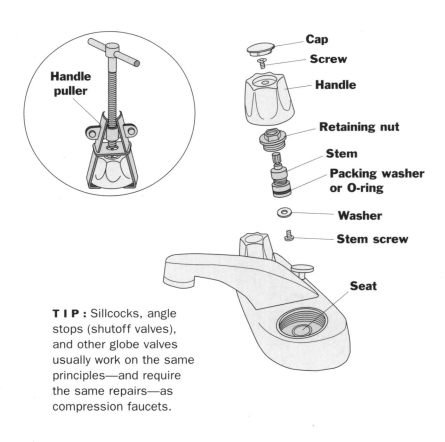

Handle puller

Cap
Screw
Handle
Retaining nut
Stem
Packing washer or O-ring
Washer
Stem screw
Seat

TIP: Sillcocks, angle stops (shutoff valves), and other globe valves usually work on the same principles—and require the same repairs—as compression faucets.

Next feel the valve seat for pits, cracks, and wear. If it is damaged, remove the seat with a valve-seat wrench or allen wrench and replace it. Reassemble the faucet. On older faucets, packing string was used in place of O-rings. If the faucet leaks only when it is turned on, loosen the packing nut and slide it out of the way. Wrap new packing string around the stem, then tighten the packing nut.

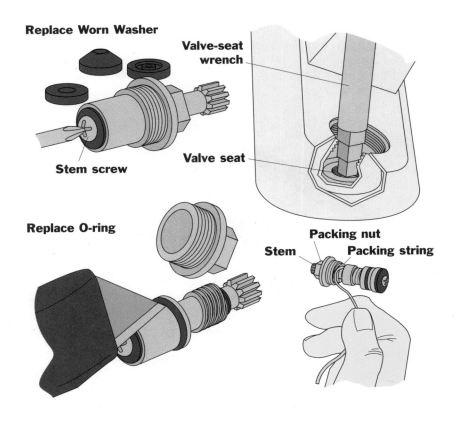

Replace Worn Washer

Valve-seat wrench

Stem screw

Valve seat

Replace O-ring

Stem

Packing nut

Packing string

Fix ball faucet

Buy a repair kit for your make and model of faucet. Turn off the water supply. Loosen the setscrew, remove the handle, and unscrew the cap. Using the tool provided in the repair kit, loosen the cam. Remove the cam, gasket, and ball. With a screwdriver or needle nose pliers, remove the seals and spring. Lift off the spout and replace the O-rings. Replace the spout and insert the new seals and spring. Reassemble the faucet, using all of the parts in the repair kit.

TIP: When using metal tools on a polished surface, protect the surface with a rag or several layers of tape.

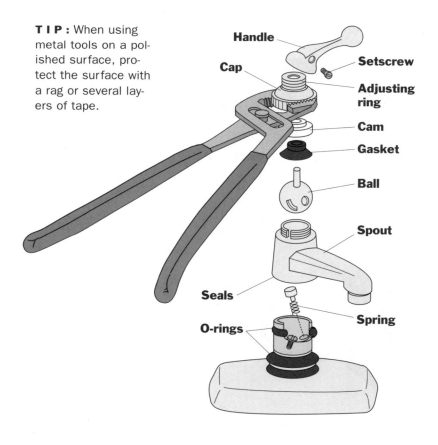

Fix cartridge faucet

Buy a repair kit, which will include a new cartridge. Turn off the water supply. Pry off the cap with a small screwdriver and remove the handle screw and handle. Remove the pivot nut or retaining clip that secures the cartridge. Then use pliers to pull the cartridge out (some kits include a puller). Install the new cartridge, aligning the tab or tabs. Replace the nut or clip. Lift off the spout and replace the O-rings. Coat the cartridge and O-rings with heat-proof grease and reassemble the faucet.

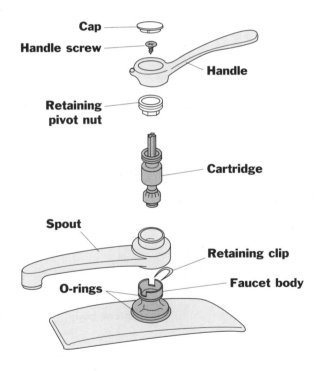

5 Fix disk faucet

Buy a repair kit for your make and model of
faucet. Turn off the water supply. Loosen the
setscrew, lift off the handle, and remove the cap.
Unscrew the disk mounting screws and lift out
the ceramic disk assembly. Remove the rubber seals on
the bottom of the disk. Clean and rinse the holes, install the
new seals, and reassemble the faucet.

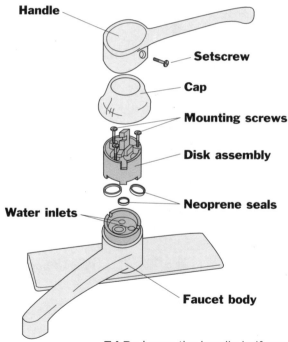

Handle

Setscrew

Cap

Mounting screws

Disk assembly

Neoprene seals

Water inlets

Faucet body

TIP: Leave the handle half open when
you turn on the water supply. Let water
flow before closing the faucet. If the leak
persists, replace the disk.

6 Replace sprayer

Shut off the water supply. Cut the hose and pull it through the sink opening. Loosen the nut under the faucet and remove the hose. Coat the male threads with pipe joint tape or compound, then screw on the new hose. Run the other end of the hose through the sink fitting, then slide on the coupling, retaining ring, and washers (or plastic rings). Apply pipe joint tape or compound to the male threads of the spray head and attach it to the coupling ring on the hose.

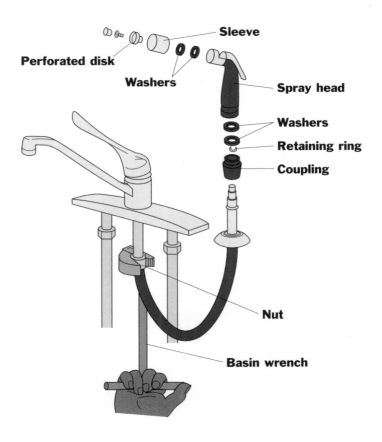

Sleeve

Perforated disk

Washers

Spray head

Washers

Retaining ring

Coupling

Nut

Basin wrench

13

7 Replace faucet

Remove the existing faucet and clean off any old putty or caulk on the sink or countertop. Apply a bead of plumber's putty or silicone caulk around the bottom of the new faucet. Set it in place and press down firmly. From below, twist the washers and mounting nuts on the threaded connections. Check that the faucet is square with the edge of the sink, then tighten the nuts with a basin wrench. Attach the supply risers first to the threaded connections on the faucets, then to the shutoff valves.

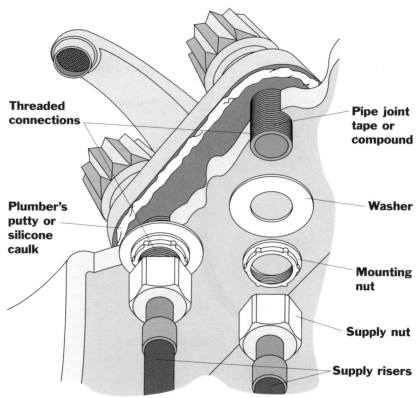

TIP: Supply risers are available in plastic, braided stainless steel, and chrome-plated brass or copper. Be sure that the nuts are the right size (⅜ inch or ½ inch) and thread style (compression or IPS).

If you are installing new chrome-plated supply risers, hand-tighten the supply nut and carefully bend each riser to align with the shutoff valve; avoid kinks. Mark the proper length, then remove and cut the risers. Attach the risers first to the faucets, then to the compression fittings on the shut-off valves. Use pipe joint tape or compound on the threaded connections. Braided supply risers cannot be cut, so buy the right length. Some assemblies may include a drain stopper (see page 21) or a spray hose (see page 13).

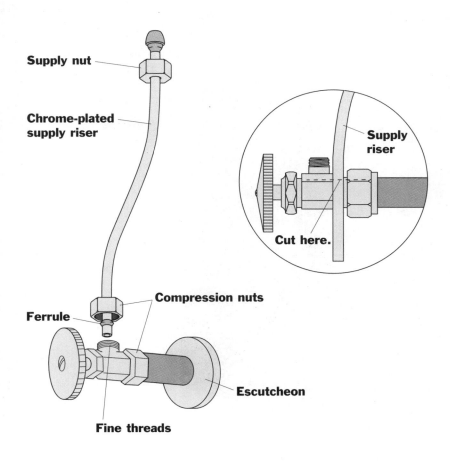

Supply nut

Chrome-plated supply riser

Supply riser

Cut here.

Compression nuts

Ferrule

Escutcheon

Fine threads

Sinks

Stop leaks

For leaks under the faucet body, tighten the mounting hardware beneath the sink. If the leak persists, remove the faucet, apply plumber's putty or silicone caulk around the bottom, and reinstall (see page 14). For leaks at the sink edge, tighten the sink mounting clips. If leaks persist, remove the sink, apply fresh putty, and reinstall (see pages 23 to 25). If the sink is glued, reseal it with clear silicone caulk. For leaks in the drain pipes, tighten all connections and replace worn parts. Check the disposer, dishwasher mounting, and drain connections.

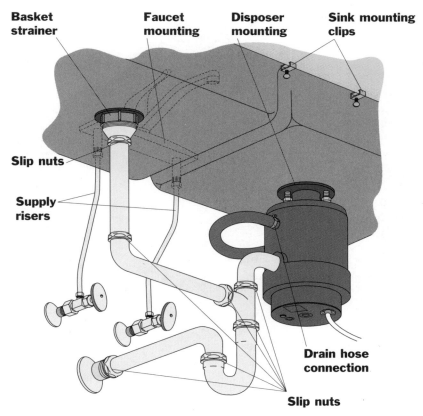

Basket strainer

Faucet mounting

Disposer mounting

Sink mounting clips

Slip nuts

Supply risers

Drain hose connection

Slip nuts

T I P : Have a helper splash water around the sink while you look for the leak underneath.

2 Unclog sink drain

If the sink has a pop-up stopper, remove it (twist it counterclockwise) and check for accumulated hair and other debris. If plunging proves necessary, use a wet rag to seal the overflow hole; also the second drain if a double sink. Lightly coat the plunger rim with petroleum jelly and place over the drain. For a tight seal, add 1 to 2 inches of water to the sink. Vigorously work the plunger up and down for two to three minutes.

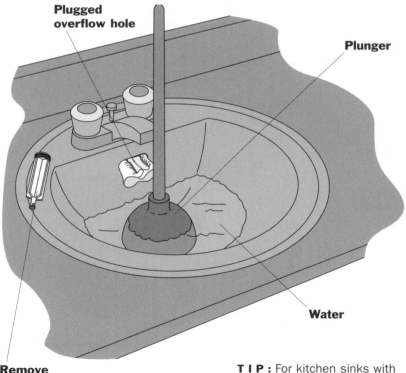

Plugged overflow hole

Plunger

Water

Remove stopper.

T I P : For kitchen sinks with a disposer hose leading to an air gap, seal the hose with two woodblocks squeezed in a C-clamp.

If plunging fails, place a container under the P-trap. If the trap has a cleanout plug, remove it and probe for clogs with a wire. If it doesn't have a cleanout, loosen the slip nuts with channel lock pliers and remove the trap. Pour out the water and remove any debris. If the P-trap isn't clogged, run a drain auger into the drainpipe. Reassemble the trap, then run hot water down the drain for a few minutes.
Use a chemical drain cleaner only as a last resort and follow directions carefully.

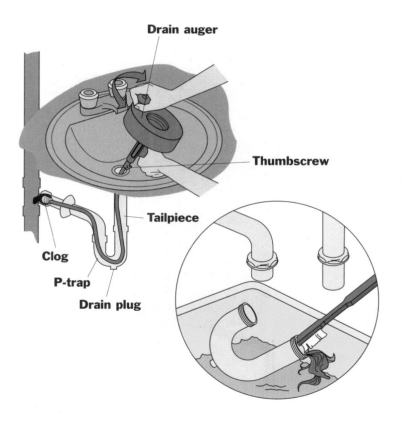

Drain auger

Thumbscrew

Tailpiece

Clog

P-trap

Drain plug

3 Free jammed disposer

Turn off electric power to the disposer. If possible, find the wrench originally supplied with the disposer. It fits into the bottom of the unit, allowing you to release the jammed object by turning the mechanism back and forth. Alternatively, insert a thick dowel down into the disposer and lever the impeller until it spins freely. Remove whatever object jammed the impeller and restore power. Then push the reset button, turn on the cold water, and let the disposer run a few minutes.

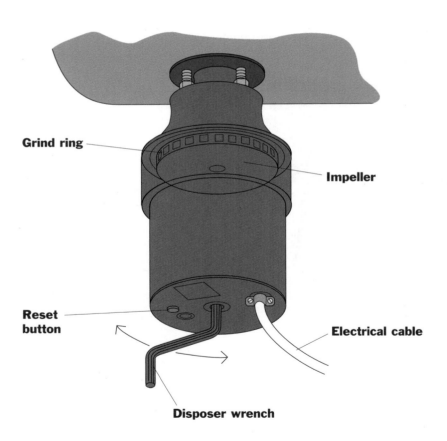

Grind ring

Impeller

Reset button

Electrical cable

Disposer wrench

4 Adjust drain stopper

Make sure that the stopper is properly engaged
with the pivot rod. Some stoppers lift out, some
twist on and off, and others require that the
pivot rod be removed first. To set the stopper
higher, loosen the screw that secures the clevis strap to the
lift rod (you may need pliers). Slide the strap farther down
the rod and tighten the screw. To set the stopper lower,
slide the clevis strap farther up the rod. Tighten or loosen
the retaining nut so that the assembly operates smoothly.

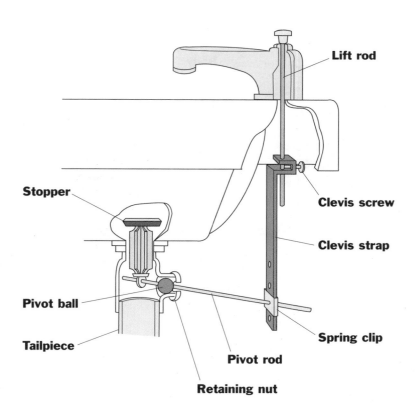

Lift rod

Stopper

Clevis screw

Clevis strap

Pivot ball

Spring clip

Tailpiece

Pivot rod

Retaining nut

5 Replace basket strainer

Unscrew the slip nuts at both ends of the tail-piece and remove it. Loosen the locknut using a spud wrench or by tapping it with a woodblock and hammer. Have a helper hold the strainer body in place from above, using pliers handles braced with a screwdriver. Remove the strainer body and scrape the old putty from the drain opening. Apply a bead of plumber's putty around the opening and press the new strainer body in place. Attach gasket, friction ring, and locknut from below. Tighten the locknut and reinstall the tailpiece.

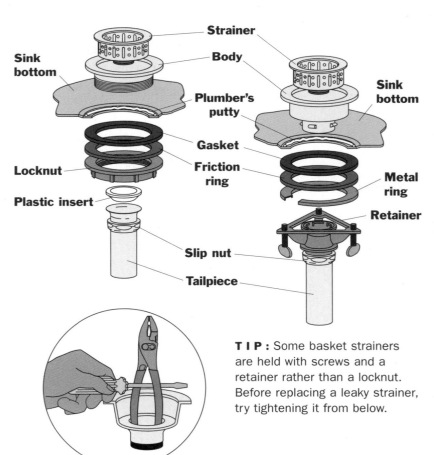

Strainer

Sink bottom

Body

Sink bottom

Plumber's putty

Gasket

Locknut

Friction ring

Metal ring

Plastic insert

Retainer

Slip nut

Tailpiece

TIP: Some basket strainers are held with screws and a retainer rather than a locknut. Before replacing a leaky strainer, try tightening it from below.

6 Remove old sink

Turn off the water supply. With a basin wrench, unscrew the mounting nuts from the threaded connections under the faucet base. Slide the supply risers out of the way. Over a bucket, loosen the slip nuts and remove the P-trap. If necessary, disconnect the sprayer, pop-up stopper, disposer, or dishwasher hookup. Remove the basket strainer and tailpiece. Use a utility knife to cut any caulk under the sink rim, remove clips or other mounting hardware, then lift out the sink.

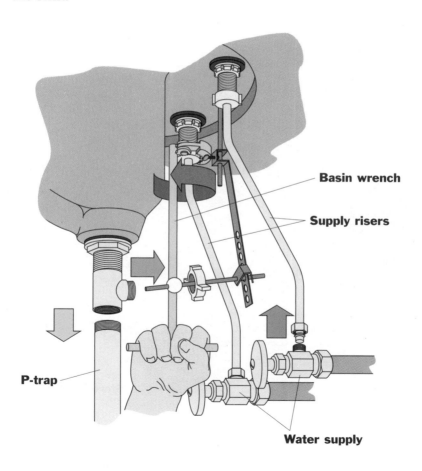

Basin wrench

Supply risers

P-trap

Water supply

7 Install countertop sink

The outer edge of a countertop ("self-rimming") sink rests on the countertop surface. To mark the sink cutout, use the paper template included with the new sink, or flip the sink upside down and trace around the edge. Mark a cut line ½ to ¾ inch inside the sink outline. The cutout should be about 1¾ inches from the front edge of the countertop. Check that the sink will rest on a flat surface all around. Drill a starter hole for the jigsaw and make the cutout.

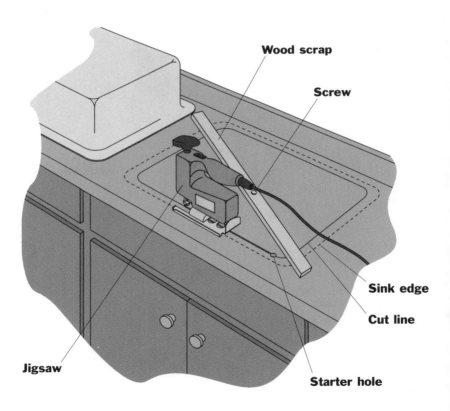

Wood scrap

Screw

Sink edge

Cut line

Jigsaw

Starter hole

Run a bead of silicone caulk on the underside of the sink rim. Turn the sink over, grasp it by the drain opening, and carefully lower it in place. Center the sink in the cutout and check that the front edge is parallel to the countertop edge. Press around the rim for a complete seal. If you are installing a stainless steel sink, install the mounting clips, included with the unit, from below.

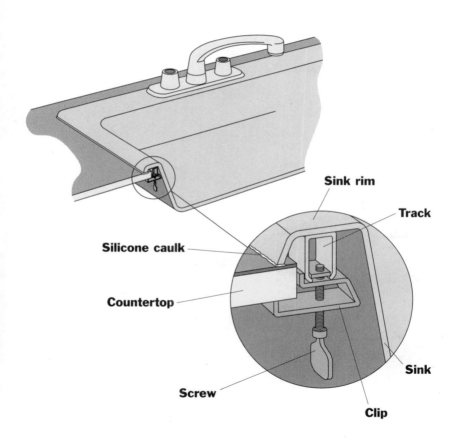

SINKS

8 Install pedestal sink

Most pedestal sinks have a wall bracket that bears the weight of the sink. For best results, install 2×6 blocking between the studs for the bracket. Set the pedestal and sink in place, centered between the shutoff valves. Make sure the sink is level. Use a pencil or masking tape to outline the pedestal on the floor and the sink on the wall. Mark the floor and wall for the mounting holes in the pedestal base and the wall bracket. Slide the assembly out of the way and drill pilot holes.

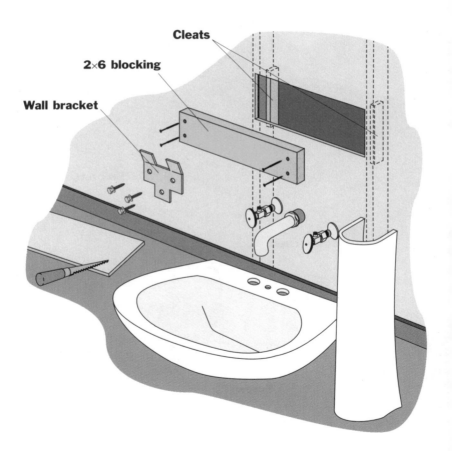

Cleats

2×6 blocking

Wall bracket

Reposition the pedestal and secure it with lag screws and washers. Install the faucet (see page 14) and pop-up stopper (see page 21). Set the sink on top of the pedestal and loosely install supply risers and drain fittings. Attach the mounting hardware to the wall and install the sink. Hook up the water supply and drain assembly. Caulk between the sink and wall.

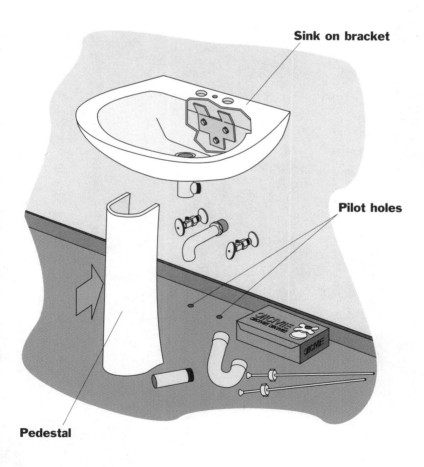

Sink on bracket

Pilot holes

Pedestal

Tubs and Showers

1 Repair tub leak

To repair a leak between the tub and waste arm, first remove the drain stopper. Insert a strainer wrench in the drain and loosen the fitting. Remove the fitting and, from below, replace the gasket. From above, apply pipe joint tape or compound to the threads and put plumber's putty under the flange of the fitting before replacing it. If the drain tee leaks, loosen the slip nuts and take the unit apart. Replace the slip-nut washers. Apply pipe joint tape or compound to the threads and reassemble the tee joint.

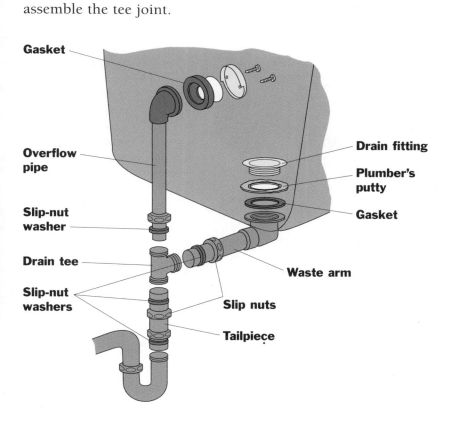

Gasket

Overflow pipe

Slip-nut washer

Drain tee

Slip-nut washers

Drain fitting

Plumber's putty

Gasket

Waste arm

Slip nuts

Tailpiece

2 Unclog tub drain

First try a plunger (see page 18). If this fails, run
a drain auger down the overflow pipe. To gain
access, remove the stopper linkage by unscrewing
the overflow plate and then pulling out the
assembly (opposite). Remove hair or other debris from the
plunger and linkage. Then work the auger into the drainpipe
until it contacts the blockage. Tighten the screw on the
handle, then rotate the handle back and forth while pushing
down. Repeat until the clog is loosened. Run water down
the drain for several minutes.

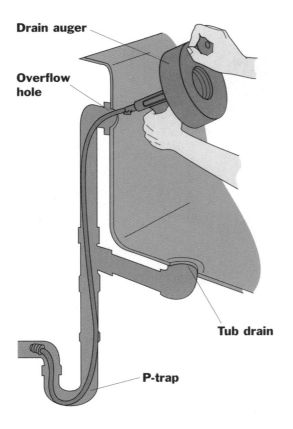

Drain auger

**Overflow
hole**

Tub drain

P-trap

T I P : If other fix-
tures are clogged,
the obstruction is
farther down a
drain line. Find the
cleanout and try to
reach the clog with
an auger.

3 Adjust drain stopper

To fix a leaky trip-lever drain with a concealed stopper, try lengthening the linkage. Remove the overflow plate and take out the whole assembly. Loosen the locknut and turn the threaded rod. Make a small adjustment, then reassemble and test. Repeat the process until the stopper holds water. For a pop-up stopper, open the drain and pull out the stopper and linkage. Replace the O-ring if it appears worn. If the stopper isn't dropping far enough into the drain, remove the overflow plate and adjust the linkage.

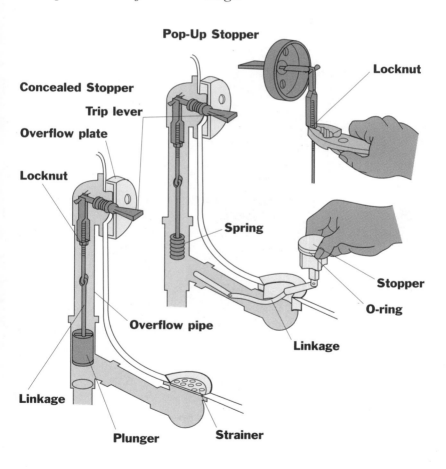

Pop-Up Stopper

Concealed Stopper

Trip lever

Overflow plate

Locknut

Locknut

Spring

Stopper

O-ring

Overflow pipe

Linkage

Linkage

Plunger   Strainer

Repair shower valve

Most shower and tub valves can be repaired like sink faucets. Remove the handle and escutcheon. Some single-handled shower valves have built-in shutoff valves. You can turn off the water supply at these valves or at the main supply valve for the house. Remove the bonnet nut or retaining ring, then grasp the cartridge with pliers and pull it out. Replace any O-rings. If the leak persists, replace the cartridge.

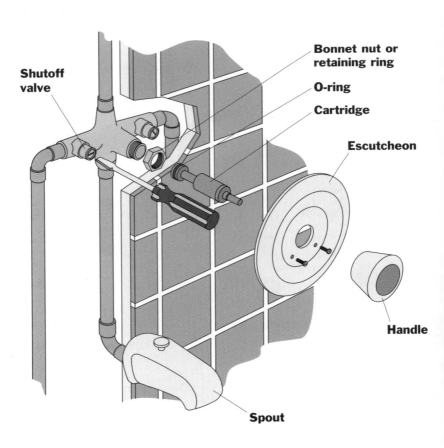

Shutoff valve

Bonnet nut or retaining ring

O-ring

Cartridge

Escutcheon

Handle

Spout

Repair diverter valve

Turn off the water supply. Remove the diverter valve handle (the screw may be under a cap) and escutcheon. With a deep socket wrench, carefully loosen and remove the bonnet nut. You can repair a compression diverter much like a compression faucet (see pages 8 and 9). Disassemble the valve and replace washers, O-rings, or packing. If the valve is too worn, or if it is a cartridge style, replace it.

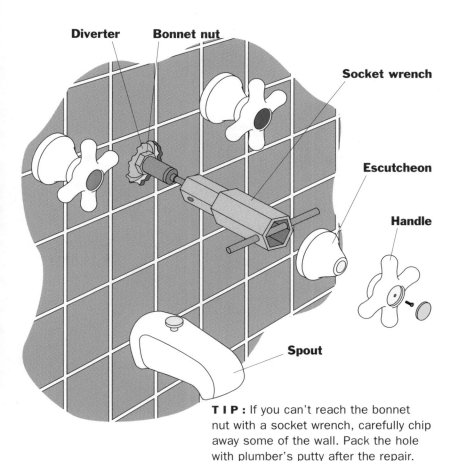

Diverter   Bonnet nut

Socket wrench

Escutcheon

Handle

Spout

T I P : If you can't reach the bonnet nut with a socket wrench, carefully chip away some of the wall. Pack the hole with plumber's putty after the repair.

6 Replace tub spout

If water drips from the spout when the shower is on, remove the spout and check the diverter mechanism. To remove the spout, insert the handle of a hammer, wrench, or large screwdriver into the spout, then turn. Remove any debris from the diverter. If this doesn't fix the problem, buy a new spout. Test the fit of the replacement spout; you may need to change nipples or add a reducing bushing. Apply pipe joint tape or compound to the nipple threads, then screw on the spout.

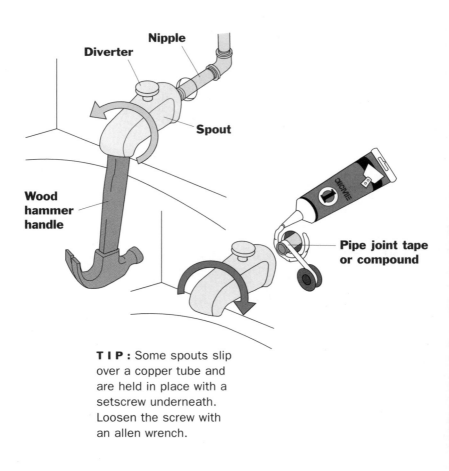

Diverter   **Nipple**

Spout

Wood hammer handle

Pipe joint tape or compound

T I P : Some spouts slip over a copper tube and are held in place with a setscrew underneath. Loosen the screw with an allen wrench.

7 Repair showerhead

If water drips from the showerhead when the water is off, the valve (faucet) needs repair. If water leaks around the threads on the shower arm, remove the showerhead and replace any washers and O-rings. Apply pipe joint tape or compound to the threads on the shower arm and replace the showerhead. To clear a clog, remove and disassemble the showerhead. Rinse loose debris and clean the water holes with a pin. Remove mineral deposits by soaking the parts overnight in a vinegar-and-water solution. Reassemble the showerhead.

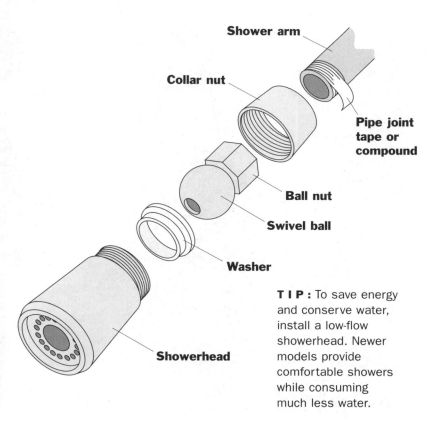

Shower arm

Collar nut

Pipe joint tape or compound

Ball nut

Swivel ball

Washer

Showerhead

T I P : To save energy and conserve water, install a low-flow showerhead. Newer models provide comfortable showers while consuming much less water.

Toilets

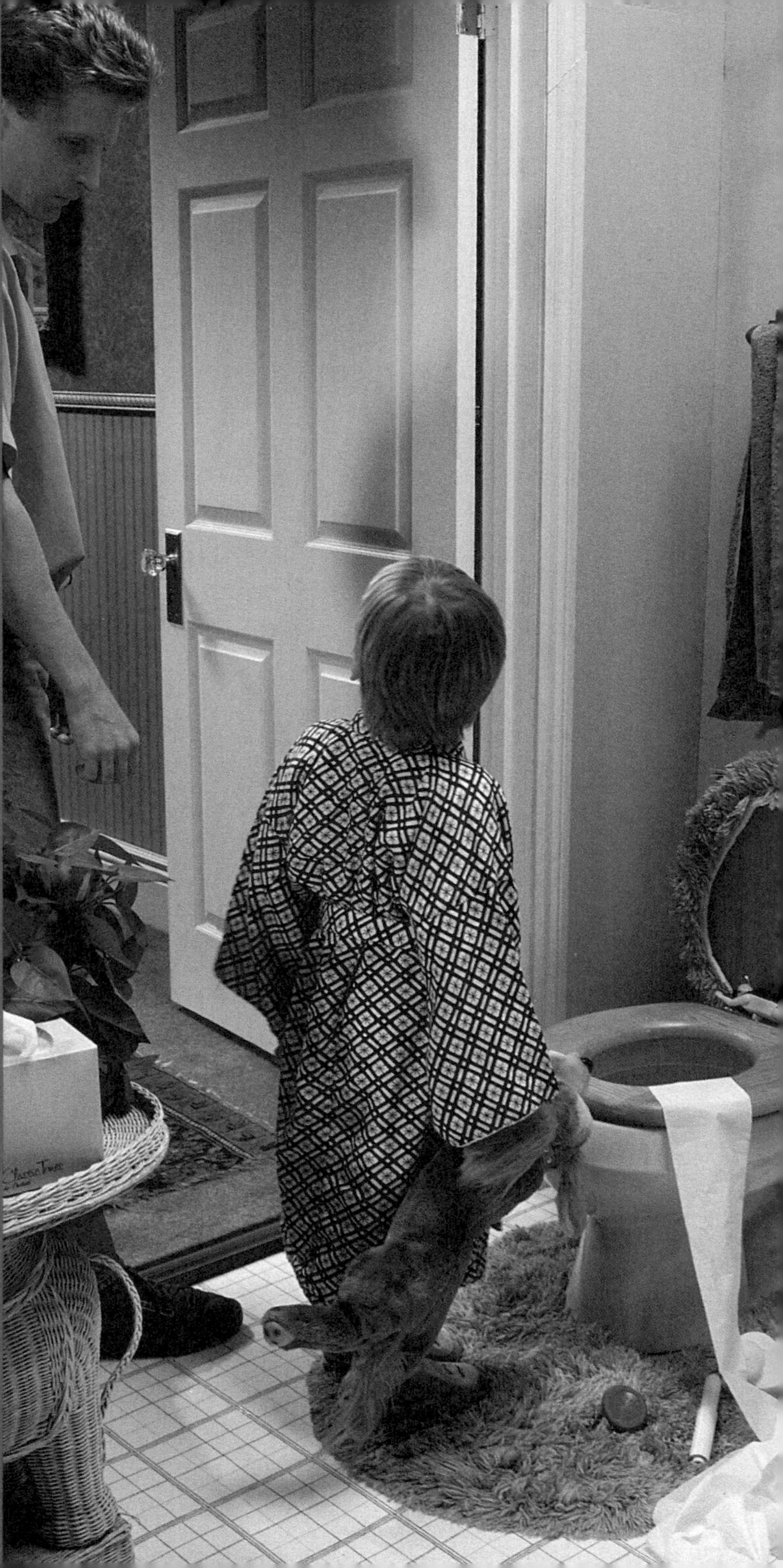

Unclog toilet

Use a flanged plunger for several minutes. If the problem persists, snake a closet auger down the bowl. Retract the cable into the shaft by pulling up on the handle, then insert the shaft into the bowl. Push the handle down to work the cable into the drain until it contacts the clog. Rock the handle back and forth gently while pushing and pulling slightly. Be patient—it may take time to break up the clog.

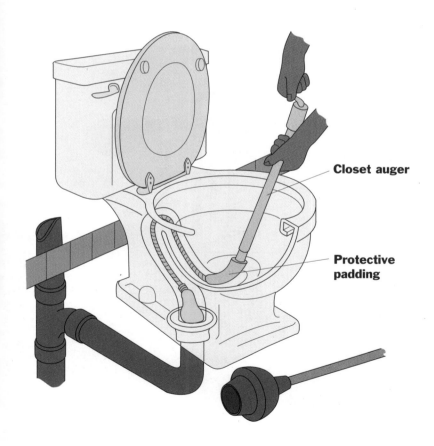

Closet auger

Protective padding

Stop toilet running

Tap the handle and check the adjustments on the lift chain or lift rods. If water is flowing into the overflow tube, adjust the water level in the tank. For a toilet with a float ball, carefully bend the arm in the middle to force the float down. After the toilet flushes, the water level should rest ½ inch below the overflow tube. For a toilet with a floating-cup ball cock, squeeze the clip on the side to adjust the cup. If the water level is too low, reverse the adjustments.

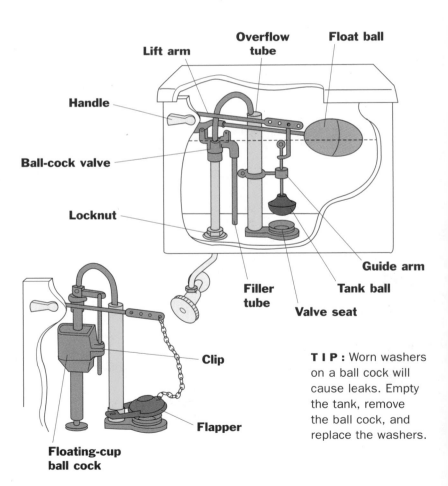

T I P : Worn washers on a ball cock will cause leaks. Empty the tank, remove the ball cock, and replace the washers.

Stop toilet leaking

Water on the floor beneath the tank often results from condensation. Improving ventilation may help. You can also install an insulated tank liner.

If the problem persists, tighten—but don't over-tighten—the nuts on the closet bolts and tank bolts. If water is leaking around the base of the toilet, the wax ring may need replacing (see pages 40 and 41). Check the water level in the tank and tighten water supply connections. Examine the ball-cock valve for a leak. Look for cracks in the tank or bowl.

Where Leaks Occur

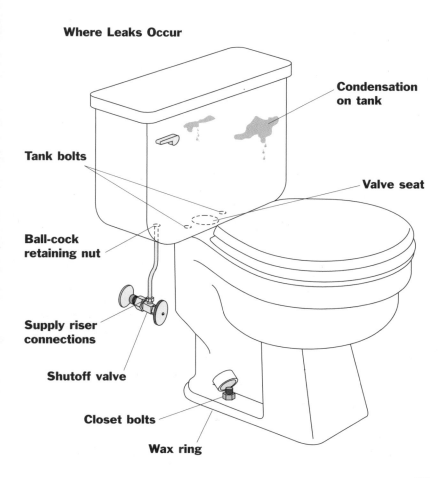

Condensation on tank

Tank bolts

Valve seat

Ball-cock retaining nut

Supply riser connections

Shutoff valve

Closet bolts

Wax ring

4 Replace toilet

Shut off the water supply and empty the water from the tank and bowl. Disconnect the supply riser and remove the nuts holding the tank to the bowl. Lift the tank off the bowl. Remove the nuts at the bottom of the bowl and lift the bowl off the closet flange. Plug the flange with a rag to prevent sewer gas from entering the bathroom. Clean off the flange with a putty knife and replace the closet bolts, positioning them upright with dabs of plumber's putty.

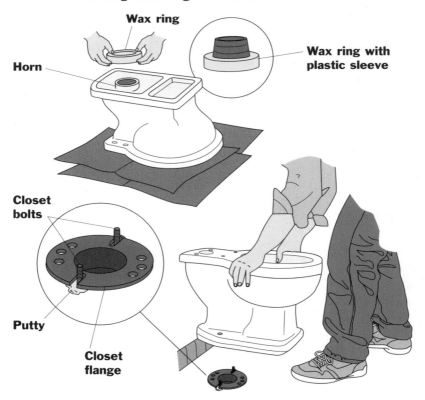

Setting Wax Ring Onto Horn

Wax ring

Horn

Wax ring with plastic sleeve

Closet bolts

Putty

Closet flange

Setting Bowl On Closet Flange

With the bowl upside down, press a new wax ring onto the horn. Remove the rag from the flange. Turn the bowl over and lower it onto the flange, using the closet bolts as a guide. Push down on the bowl, then install the washers and nuts. Tighten each side a little at a time; don't overtighten. Set the gasket and rubber cushions in place on the bowl and lower the tank onto them. Install the tank bolts, washers, and nuts, tightening them carefully. Reattach the water supply and install the seat.

Attaching the Tank

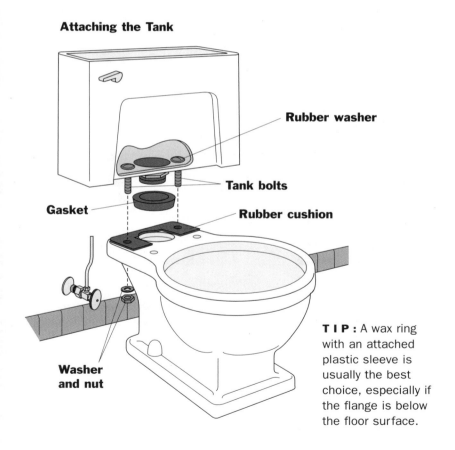

Rubber washer

Tank bolts

Gasket

Rubber cushion

Washer and nut

T I P : A wax ring with an attached plastic sleeve is usually the best choice, especially if the flange is below the floor surface.

5 Replace ball cock

Turn off the water supply. Flush the toilet, then sponge out the remaining water. Loosen the nut underneath the tank and remove the old ball cock. Install the new ball cock according to manufacturer's instructions. Connect and turn on the water supply. Flush a few times to check for leaks. Adjust the ball cock so that the water level stabilizes ½ inch below the top of the overflow tube. Since the tank is drained, you may want to go ahead and replace the flush valve as well.

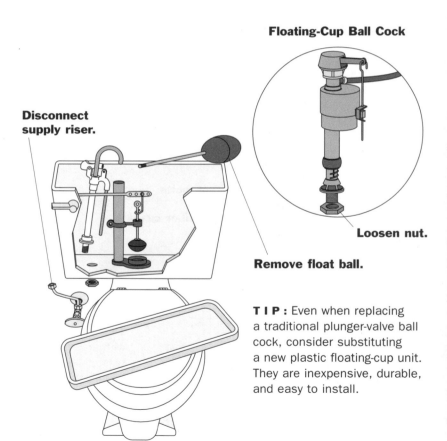

Floating-Cup Ball Cock

Disconnect supply riser.

Loosen nut.

Remove float ball.

T I P : Even when replacing a traditional plunger-valve ball cock, consider substituting a new plastic floating-cup unit. They are inexpensive, durable, and easy to install.

6 Replace flush valve

Water constantly trickling into the toilet bowl is probably caused by a faulty flush valve. If it can't be adjusted to seal properly, buy a replacement unit. Turn off and disconnect the water supply, then empty the tank. Remove the tank from the bowl and unscrew the spud nut with a spud wrench or large adjustable pliers. Slip the rubber washer on the new flush valve and set the valve in the tank. Replace the spud nut. Set the spud washer on the valve tailpiece before reinstalling the tank.

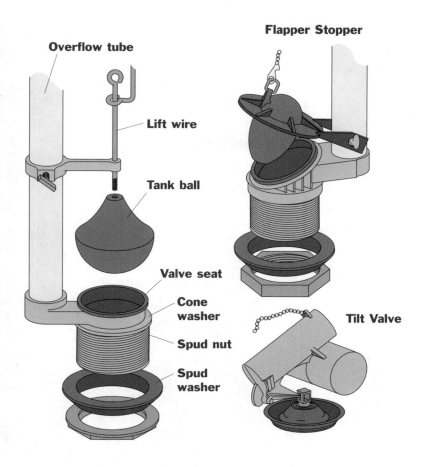

Overflow tube

Flapper Stopper

Lift wire

Tank ball

Valve seat

Cone washer

Spud nut

Spud washer

Tilt Valve

Water Heaters

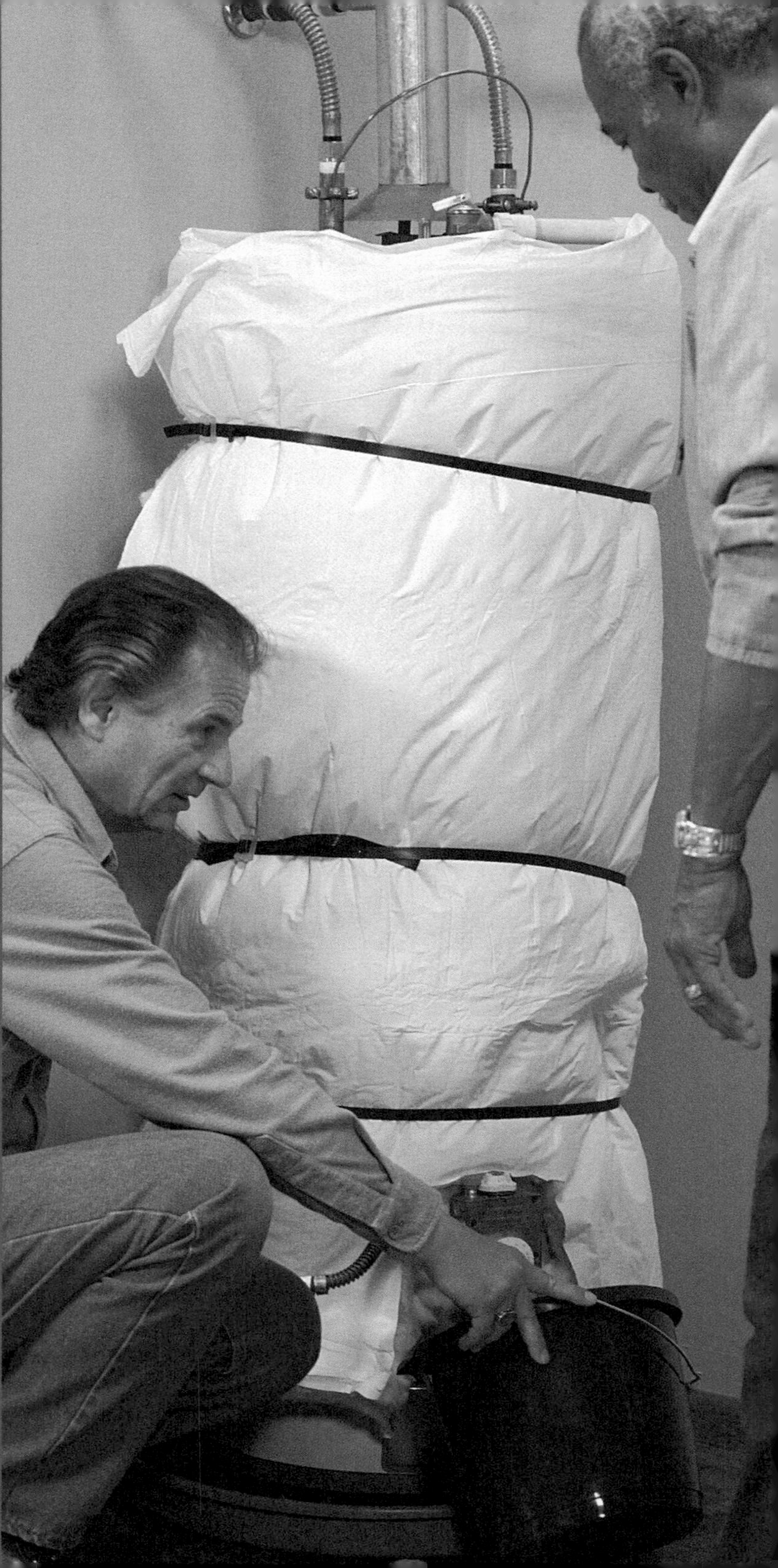

Maintain regularly

Drain 1 or 2 gallons from the tank every month to prevent sediment accumulation. Empty the tank annually using a garden hose, then turn on the inlet valve and let water run through the tank and out the hose. Every six months, test the temperature-pressure relief valve (TPRV) by lifting the valve lever to let water run out. If it fails to open, or leaks when it's closed, replace it. Inspect the anode every one to four years, depending on water characteristics. Replace the anode if worn.

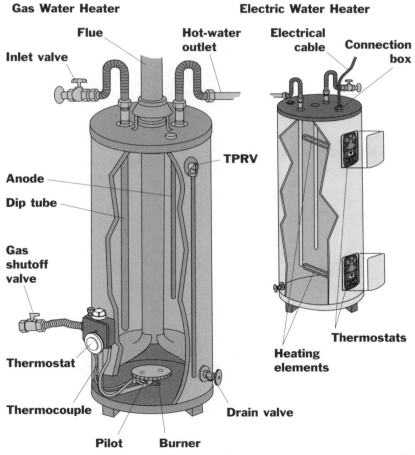

Gas Water Heater

Inlet valve
Flue
Hot-water outlet
TPRV
Anode
Dip tube
Gas shutoff valve
Thermostat
Thermocouple
Pilot
Burner
Drain valve

Electric Water Heater

Electrical cable
Connection box
Thermostats
Heating elements

2

Relight pilot (gas water heater)

Make sure all gas supply valves are open. Set the gas control to *pilot*. While pressing the pilot button, light the pilot. Let it burn for a minute or two, release the pilot button, and turn the control to *on*. If the pilot won't stay lit, turn off the gas supply, disconnect the pilot gas line, and run a thin wire through it. Check for drafts along the floor that may extinguish the pilot. If the pilot lights but goes out after releasing the pilot button, the thermocouple may need adjustment or replacement.

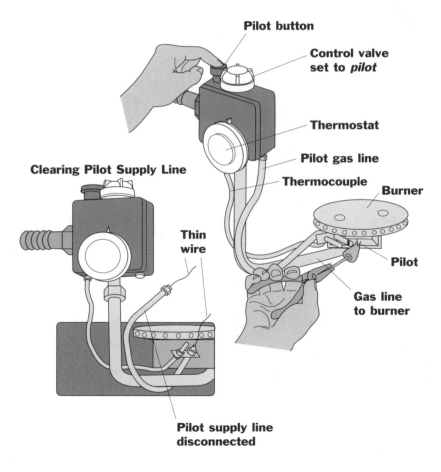

Pilot button

Control valve set to *pilot*

Thermostat

Pilot gas line

Thermocouple

Burner

Clearing Pilot Supply Line

Thin wire

Pilot

Gas line to burner

Pilot supply line disconnected

3 Replace thermocouple (gas water heater)

The thermocouple, which cuts off the gas if it detects that the flame has gone out, can wear out. Replace the thermocouple by first turning off the gas supply. With two wrenches, unscrew the nuts that hold the thermocouple to the burner bracket. (You may have to remove the burner assembly first.) Disconnect the thermocouple from the thermostat and remove it. Install the new thermocouple. Light the pilot and make sure the tip of the thermocouple is surrounded by flame.

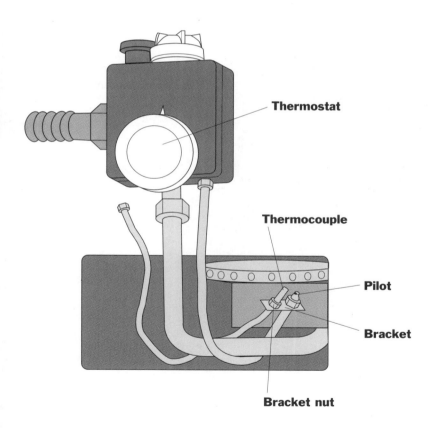

Thermostat

Thermocouple

Pilot

Bracket

Bracket nut

Replace dip tube

The dip tube delivers cold water to the bottom of the water heater. If it is worn or damaged, incoming cold water chills the hot water toward the top of the heater. To replace this inexpensive component, close the inlet valve and drain off a small bucket of water from the drain valve. Detach the cold-water supply connector and remove the nipple. Slip a wood dowel in the dip tube and twist it out. Install a new one and reconnect the water supply. Use pipe joint tape or compound on male threads.

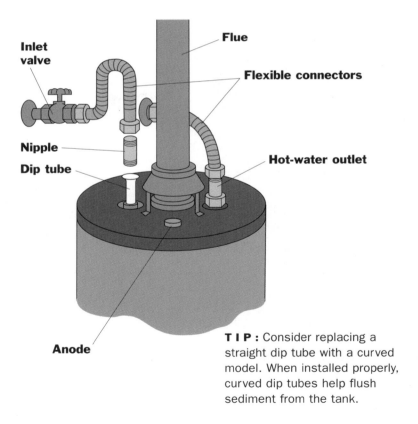

Inlet valve

Flue

Flexible connectors

Nipple

Dip tube

Hot-water outlet

Anode

TIP: Consider replacing a straight dip tube with a curved model. When installed properly, curved dip tubes help flush sediment from the tank.

Replace electric heating element

Turn off the electric power to the water heater. If the heater has two heating elements, you'll need to use a volt-ohmmeter to determine which one needs replacement, or simply replace both. With the power still off, disconnect the wires, turn off the water supply, and drain the tank. Unscrew the nut on the heating element (or remove the mounting-flange bolts), and ease the element out. Buy and install an identical replacement. Connect the wires, set the temperature, restore the power and water supply, and press the reset button.

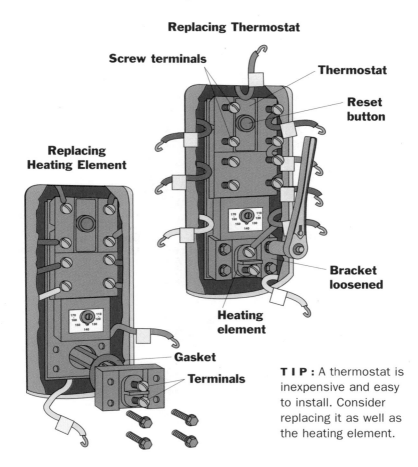

Replacing Thermostat

Screw terminals

Thermostat

Reset button

Replacing Heating Element

Bracket loosened

Heating element

Gasket

Terminals

TIP : A thermostat is inexpensive and easy to install. Consider replacing it as well as the heating element.

Remove water heater

Puddling at the base of a heater may indicate a leaky inlet or outlet pipe connection. But if the tank itself is leaking, you'll have to replace the water heater. If you're installing a similar tank, you can do the job yourself. If you are switching from electric to gas, however, or want a much larger or smaller tank, hire a professional. Remove the heater by first closing the gas valve or turning off the electric power. Attach a garden hose to the drain valve, open a hot-water faucet, and drain the tank.

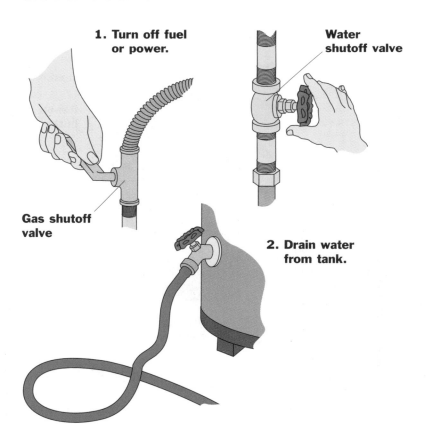

1. Turn off fuel or power.

Water shutoff valve

Gas shutoff valve

2. Drain water from tank.

If replacing a gas water heater, disconnect the gas line and flue piping. For an electric unit, detach the cover on the connection box, remove the wire nuts, and disconnect the wires. If there are flexible connectors on the supply lines, unscrew them. If the water supply pipes are rigid copper, cut the cold-water pipe a couple of inches below the valve. Then cut the hot-water pipe to the same length. Solder a male adapter onto each pipe (see pages 56 and 57). Remove the water heater.

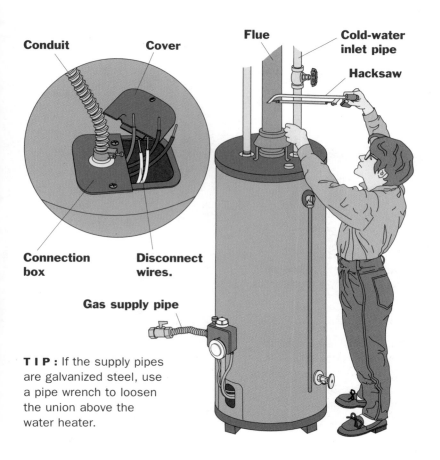

Conduit   Cover   Flue   Cold-water inlet pipe

Hacksaw

Connection box   Disconnect wires.

Gas supply pipe

T I P : If the supply pipes are galvanized steel, use a pipe wrench to loosen the union above the water heater.

7 Install water heater

Set the water heater in place. Use shims if necessary to keep it plumb. Install threaded brass nipples into the inlet and outlet openings on the water heater. Install flexible connectors between the nipples and the water pipes. Screw a new temperature-pressure relief valve (TPRV) into the designated hole in the top or side of the water heater. (Make sure it has the right pressure and BTU rating.) Attach the old discharge pipe to the TPRV. Apply pipe joint tape or compound to all male threads before assembling.

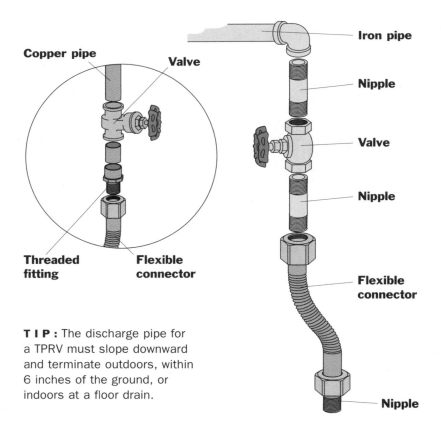

Copper pipe

Valve

Iron pipe

Nipple

Valve

Nipple

Threaded fitting

Flexible connector

Flexible connector

Nipple

TIP: The discharge pipe for a TPRV must slope downward and terminate outdoors, within 6 inches of the ground, or indoors at a floor drain.

On a gas water heater, attach the flue hat and reconnect the flue. Secure all flue joints with sheet-metal screws. With a direct-vent model, follow the manufacturer's instructions. Hook up the gas supply (if you need new piping, contact a professional). On an electric water heater, remove the cover on the connection box. Hook up the metal conduit to the box connector, then connect the wires with wire nuts. Connect the green wire to the grounding terminal in the box. Fill the tank before restoring power. Adjust the temperature.

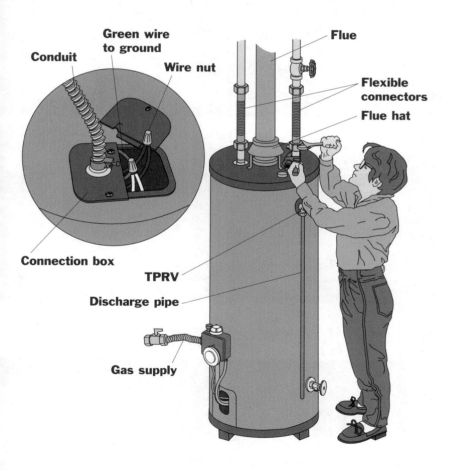

FIPS 1" 90°
(5) (15)

F
(20

1 x ¾ x ¾ TEES 1" ST. 90°
(6) 1 (10)

DR
(6) ·

1 x ½ BUSH. 1" 45°
1 x ½ RED. 1 (10)

½ x
(10)

1 x ½ TEES ST. 45°
(6) (10)

¾
·

1 x ¾ TEES TEES
(4) (5)

¾

Threaded connections

Iron pipe, whether precut and threaded or cut and threaded to order, is available from most hardware stores. For repairs, you need a pair of pipe wrenches to loosen and tighten joints. Spread pipe joint tape or compound on the male threads before joining. Threaded fittings are also used when joining plastic to metal pipe, or copper to iron pipe. Generally, a plastic male fitting should be used with a metal female fitting. These connections also require the use of pipe joint tape or a compound compatible with the plastic.

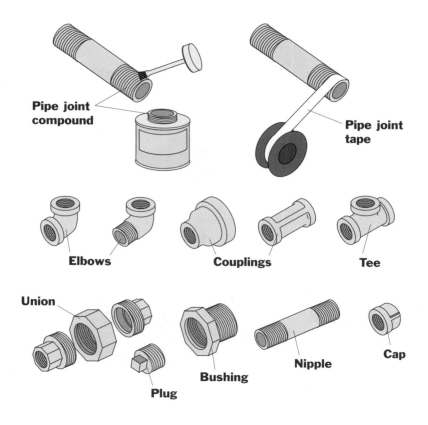

Pipe joint compound

Pipe joint tape

Elbows

Couplings

Tee

Union

Plug

Bushing

Nipple

Cap

2 Join copper pipe

Cut the pipe with a tubing cutter. Rotate the cutter while gently tightening the handle. Use the reamer on the pipe cutter to remove the burr on the inside edge of the cut pipe. Using sand cloth, shine the outside end of the pipe. Shine the inside with sand cloth or a properly sized fitting brush. With a flux brush, apply a thin layer of flux to the pipe end. Place the fitting on the pipe and twist to spread the flux.

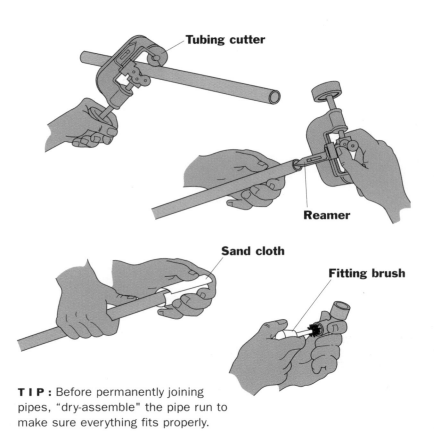

TIP: Before permanently joining pipes, "dry-assemble" the pipe run to make sure everything fits properly.

Unwind about 8 inches of lead-free solder and bend the last inch 90 degrees. Light a propane torch with a match or spark lighter and adjust the flame. Heat the fitting only, on all sides, until the flux begins to sizzle. Touch the solder to the joint. When it melts, quickly remove the torch and push the solder around the joint. Use ½ to ¾ inch of solder at each joint. Wipe off excess solder with a damp rag, but don't stress the joint.

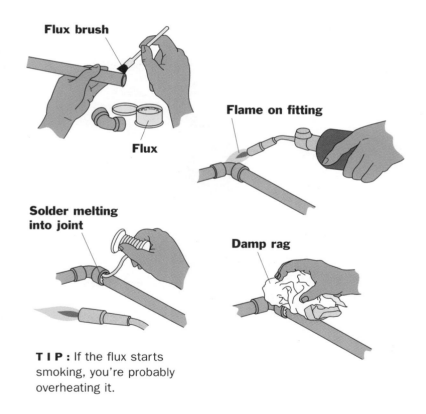

Flux brush

Flux

Flame on fitting

Solder melting into joint

Damp rag

T I P : If the flux starts smoking, you're probably overheating it.

3

Join plastic pipe

To determine the length of pipe needed, measure between the fittings, from inside of hub to inside of hub. Subtract ⅛ inch per hub. Cut the pipe with a plastic pipe cutter or a carpenter's saw, hacksaw, or chop saw (power miter saw). The cuts must be straight. Use a utility knife to scrape away burrs from the inside and outside edges. Dry-fit the entire assembly, marking the alignment on the pipe. Without cement, ABS (black) pipe won't slip all the way into a hub, so make allowances. Pull the pieces apart and begin gluing, one joint at a time.

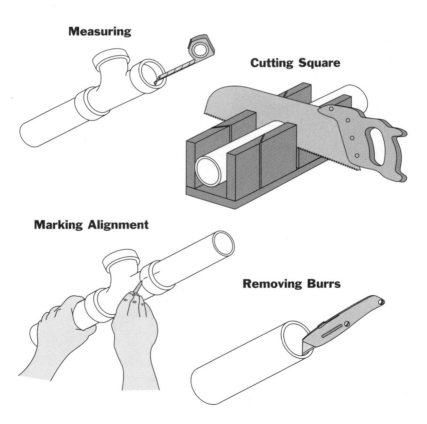

Measuring

Cutting Square

Marking Alignment

Removing Burrs

When joining PVC pipe (white or beige), coat each mating surface first with primer, then with solvent cement (both specified for PVC). ABS pipe doesn't require primer; just use ABS solvent cement. Slip the pipe into the fitting with the alignment marks one-quarter turn apart. Quickly twist into proper alignment and depth. Hold it until the cement sets; this can be a few seconds with ABS and several minutes with large-diameter PVC. Wipe excess cement with a rag.

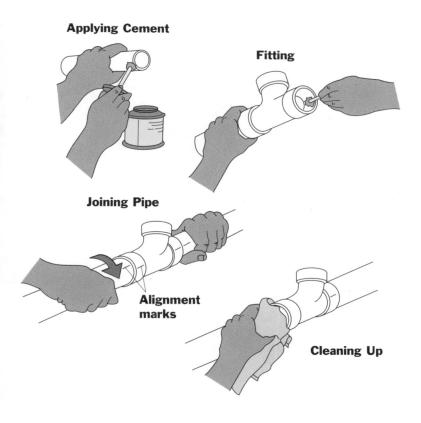

Applying Cement

Fitting

Joining Pipe

Alignment marks

Cleaning Up

Repair leaks

Turn off the water supply. To temporarily plug a small leak, jam a toothpick into it, then tighten a piece of rubber over it with a hose clamp or wrap the pipe with electrical tape. Another quick fix is a repair clamp. For a more permanent repair of iron or plastic pipe, install a dresser coupling. Cut through the damaged section, slip on the coupling components, then tighten with pipe wrenches. For a permanent repair, and for any leaking fitting, replace the pipe or fitting.

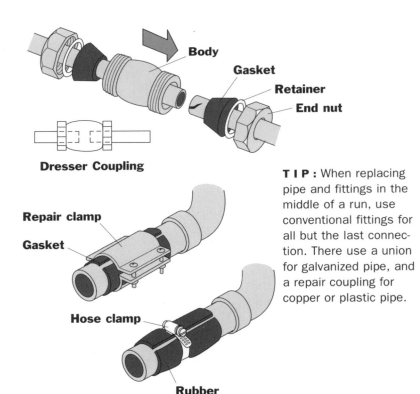

Body

Gasket

Retainer

End nut

Dresser Coupling

Repair clamp

Gasket

Hose clamp

Rubber

TIP: When replacing pipe and fittings in the middle of a run, use conventional fittings for all but the last connection. There use a union for galvanized pipe, and a repair coupling for copper or plastic pipe.

Silence noisy pipes

To reduce a loud hammering noise ("water hammer") when faucets are turned off, first try to insulate pipes from framing with insulated straps and hangers, or by packing insulation between pipes and wood. If the problem persists, install air chambers at each offending fixture. Buy water hammer arresters or make air chambers by installing lengths of pipe one size larger than the supply pipes you are muffling. They should be 12 to 18 inches long.

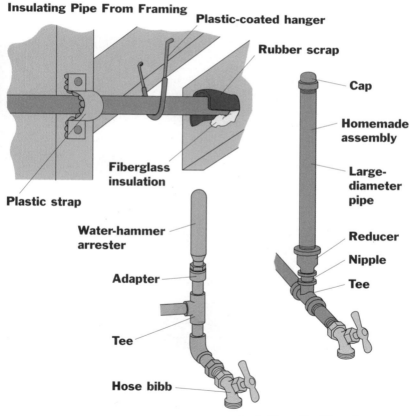

Insulating Pipe From Framing
Plastic-coated hanger
Rubber scrap
Cap
Homemade assembly
Large-diameter pipe
Fiberglass insulation
Plastic strap
Water-hammer arrester
Reducer
Nipple
Adapter
Tee
Tee
Hose bibb
Installing Air Chamber or Shock Absorber

6 Thaw frozen pipes

Shut off water supply to the pipe. Open the nearest faucet. With a hair dryer—or, on metal pipe only, a propane torch—apply heat to the pipe. Start at the faucet and work back. (Don't overheat the pipe, and don't use a propane torch near wood.) Alternatively, wrap the pipe with electrical heating tape. Once the ice has melted, turn on the water supply, then close the faucet after water runs freely. If freezing persists, insulate the pipe or permanently wrap it with electrical heating tape.

Thawing a Frozen Pipe

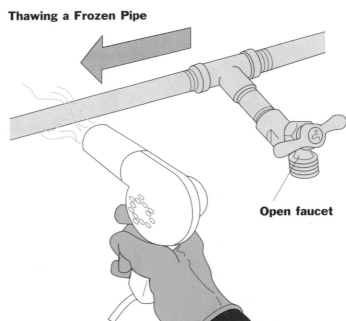

Open faucet

TIP: If a vent pipe ices over at the opening above the roof, remove the ice by running a drain auger up the vent, or by getting on the roof and clearing the pipe with a stick.

Choosing the Right Pipe Size

The following dimensions are the minimum recommended sizes required by most municipalities. Check with the building department for local requirements.

Fixture	Minimum Supply Pipe	Minimum Trap/Trap Arm Size
Bathtub	½"	1½"
Washing machine	½"	2"
Dishwasher	½"	1½"
Kitchen sink	½"	1½"
Laundry sink	½"	1½"
Shower	½"	2"
Hose bibb	½"	NA
Bar sink	⅜"	1½"
Bidet	⅜"	1¼"
Lavatory	⅜"	1¼"
Toilet	⅜"	3" (varies by fixture)

Rough-In Dimensions for Bathroom Fixtures

	Toilet	Basin	Bathtub	Shower
Distance of below-floor drainpipe from back wall	Varies; 12"	—	6–10"	Center of stall
Height of drain stub above floor	—	15–17"	—	—
Height of supply shutoff above floor	5–10"	19–21"	—	—
Distance of shutoff from centerline of fixture	6"	8"	—	—
Height of faucet above finish floor	—	—	26"	46"
Height of tub spout above finish floor	—	—	20"	—
Height of showerhead above finish floor	—	—	65–76"	65–76"

U.S./Metric Measure Conversions

Formulas for Exact Measures

	Symbol	When you know:	Multiply by:	To find:
Mass (Weight)	oz	ounces	28.35	grams
	lb	pounds	0.45	kilograms
	g	grams	0.035	ounces
	kg	kilograms	2.2	pounds
Volume	pt	pints	0.47	liters
	qt	quarts	0.95	liters
	gal	gallons	3.785	liters
	ml	milliliters	0.034	fluid ounces
Length	in	inches	2.54	centimeters
	ft	feet	30.48	centimeters
	yd	yards	0.9144	meters
	mi	miles	1.609	kilometers
	km	kilometers	0.621	miles
	m	meters	1.094	yards
	cm	centimeters	0.39	inches
Temperature	°F	Fahrenheit	$5/9$ (after subtracting 32)	Celsius
	°C	Celsius	$9/5$ (then add 32)	Fahrenheit
Area	in^2	square inches	6.452	square centimeters
	ft^2	square feet	929.0	square centimeters
	yd^2	square yards	8361.0	square centimeters
	a	acres	0.4047	hectares

Rounded Measures for Quick Reference

1 oz	=	30 g
4 oz	=	115 g
8 oz	=	225 g
16 oz	= 1 lb	450 g
32 oz	= 2 lb	900 g
36 oz	= $2^1/_4$ lb	1000 g (1 kg)
1 c	=	250 ml
2 c (1pt)	=	500 ml
4 c (1 qt)	=	1 liter
4 qt (1 gal)	=	$3^3/_4$ liter
$^3/_8$ in	=	1.0 cm
1 in	=	2.5 cm
2 in	=	5.0 cm
$2^1/_2$ in	=	6.5 cm
12 in (1 ft)	=	30.0 cm
1 yd	=	90.0 cm
100 ft	=	30.0 m
1 mi	=	1.6 km
32° F	=	0° C
212° F	=	100° C
1 in^2	=	6.5 cm^2
1 ft^2	=	930 cm^2
1 yd^2	=	8360 cm^2
1 a	=	4050 m^2